D0874287

LIQUID CRYSTALLINITY
IN POLYMERS

LIQUID CRYSTALLINITY IN POLYMERS:
Principles and Fundamental Properties

edited by Alberto Ciferri

College of Textiles, Box 8302
North Carolina State University
Raleigh, NC 27695

Dedication

Professor William R. Krigbaum was recently forced to retire due to serious illness but still managed to contribute to Chapter 2. The Authors are dedicating this book to him in recognition of his many contributions and of a life-long commitment to science.

Alberto Ciferri
College of Textiles, Box 8302
North Carolina State University
Raleigh, NC 27695

Library of Congress Cataloging-in-Publication Data

Liquid crystallinity in polymers: principles and fundamental
  properties/edited by Alberto Ciferri.

    p. cm.
  Includes bibliographical references and index.
  ISBN 0-89573-771-X
  1. Polymer liquid crystals. I. Ciferri, A.
QD923.L535   1991
530.4'29—dc20   91-7856
                                  CIP

British Library Cataloguing in Publication Data

Ciferri, A.
  Liquid crystallinity in polymers.
  1. Polymers
  I. Title
  547.7

  ISBN 0-89573-771-X

Printed in the United States of America.
ISBN 0-89573-771-X VCH Publishers
ISBN 3-527-27922-9 VCH Verlagsgesellschaft

Printing History:
10 9 8 7 6 5 4 3 2 1

Published jointly by:

VCH Publishers, Inc.   VCH Verlagsgesellschaft mbH   VCH Publishers (UK) Ltd.
220 East 23rd Street   P.O. Box 10 11 61   8 Wellington Court
Suite 909   D-6940 Weinheim   Cambridge CB1 1HW
New York, New York 10010   Federal Republic of Germany   United Kingdom

CONTENTS

LIST OF CONTRIBUTORS

Prof. Akihiro Abe
Department of Polymer Chemistry
Tokyo Institute of Technology
Ookayama, Meguro-ku
Tokyo 152
Japan

Prof. Matthias Ballauff
Polymer Institut Technische
  Universität Karlsruhe
Kaiserstrasse 12
76 Karlsruhe
Germany

Dr. Gregg L. Brelsford
Westvāco Research
Westvāco Corporation
5600 Virginia Ave.
North Charleston, S.C. 29411-2905
USA

Prof. Alberto Ciferri
Istituto di Chimica Industriale
Università di Genova
16132 Genova
Italy

Prof. Heino Finkelmann
Institut für Makromolekulare Chemie
Universität Freiburg
Stefan-Meier-Str. 31
D-7800 Freiburg
Germany

Dr. Alexei R. Khokhlov
Physics Department
Moscow State University
Moscow 117234
USSR

Prof. Maurice Kléman
Laboratoire de Physique des Solids
Université de Paris-Sud
Bâtiment 510
91405 Orsay Cédex
France

Prof. William R. Krigbaum
Department of Chemistry
Duke University
Durham, N.C. 27706
USA

Dr. Françoise Lauprêtre
ESPCI
Laboratoire de Physicochimie
Structurale et Macromoléculaire
10, rue Vauquelin
75231 Paris Cédex 05
France

Dr. Sin-Doo Lee
Optzon Systems, Inc.
3 Preston Ct.
Bedford, MA 01730
USA

Prof. Giuseppe Marrucci
Dipartimento di Ingegneria Chimica
Università di Napoli Federico II
Piazzale Tecchio
80125 Napoli
Italy

Prof. Robert B. Meyer
The Martin Fisher School of Physics
Brandeis University
Waltham, MA 02254
USA

Dr. Claudine Noël
ESPCI
Laboratoire de Physicochimie
Structurale et Macromoléculaire
10, rue Vauquelin
75231 Paris Cédex 05
France

Dr. Giorgio Ronca
Dipartimento di Chimica
Politecnico di Milano
Via Golgi 29
20131 Milano
Italy

Prof. Augusto Sirigu
Dipartimento di Chimica
Università di Napoli Federico II
Via Mezzocannone 4
80134 Napoli
Italy

Dr. Alexandra Ten Bosch
Laboratoire de Physique de la
   Matiére
Condensèe (CNRS, UA 190)
Université de Nice, Parc Valrose
06034 Nice Cédex
France

PREFACE

Ten years ago the field of liquid-crystalline polymers (LCP's) was in many respects still in its infancy. In general, researchers approached it either from the point of view of liquid crystals or from that of polymers. In the last decade, however, significant progress has been achieved and it is fair to state that liquid-crystalline polymer science has grown into a discipline in its own right, its focus being precisely the interplay between macromolecularity and liquid-crystalline order. Almost all the chapters of this book bear witness to this interrelation. The reader will note that the book is not just a collection of scattered approaches and ideas but is actually a search for correlations and for a "consensus."

Most current research is concerned with liquid crystals formed by main chain polymers in which anisometric units are aligned head to tail. It is in this area that comparison between theory and experiment can be made at a rather detailed level. However, another important field is that of sidechain liquid-crystalline polymers surveyed by Finkelmann. Here, the anisometric side groups induce liquid-crystalline order much in the usual fashion, although they have lost translational degrees of freedom because they are attached to a polymer backbone. In recent years a number of interesting new ideas have been published concerning this entropic effect. Equally important are the segmented mainchain LCP's in which the alternation of rigid and flexible segments causes an appreciable change in the orientational entropy of the chains and hence in the macroscopic properties of the material. In the chapter by Sirigu the emphasis is on their chemistry and experimental behavior because the theoretical framework for segmented liquid crystal polymers is still inadequate.

Mainchain polymers with smoothly flexing backbones have been the active center of attention for some 40 years. One can envision a first line of attack to be on the subangstrom or chemical level. Lauprêtre and Noël summarize quantum chemical calculations of the conformation of stiff chains in order to assess the basis of chain extension. On the next that is nanometer level it is often convenient to introduce the worm model characterized by one stiffness parameter, the persistence length commonly viewed as an empirical quantity. The frequently tortuous route from measurable properties of chains in dilute solution to reliable persistence lengths is outlined by Brelsford and Krigbaum. A recent trend is the assessment of chain conformation within the mesophase by simulation and neutron scattering techniques.

One obvious starting point for a theory of the polymeric nematic phase is to consider a packed system of worms interacting via their hard-core diameters. This model and several related ones are discussed by Khokhlov, who adopts the binary collision or second virial approximation valid at low volume fraction only. Other theories admit approximations whose ranges of applicability are unfortunately rather less well defined. Lattice models introduced by Abe and Ballauff, and also by Ronca, are often supposed to be reasonable for densely packed systems when the density fluctuations are small. The interest in these models arises from their versatility. A rigorous approach incorporating both hard and soft dispersion interactions is still lacking. Ten Bosch treats a semiempirical approach partly in the spirit of the molecular field introduced by Maier and Saupe.

One way of confronting these theoretical ideas with experiments is by studying detailed phase equilibria. Ciferri points out that we still have a long way to go to come up with a convincing theory of LCP's able to rationalize a huge body of data. Another interesting challenge to molecular theory is the dynamic light-scattering experiments on well-aligned samples analyzed by Lee and Meyer who extract relative values of the elastic moduli and Leslie viscosities from the time correlation functions. Measurements of the Frederiks transition allow the absolute magnitudes to be established. Their principal conclusion is that the polymer backbone behaves like a semiflexible chain.

The ratios of various mechanical properties of LCP's differ markedly from unity which is the typical value of the ratios pertaining to monomeric liquid crystals. Therefore the characteristics of the macromolecules ultimately bear on the occurrence of individual defects and texture as discussed by Kléman. He also indicates how chain ends may segregate into defect cores. I find it convenient to split up the very difficult field of nematorheology into four parts: (a) the consequences of Leslie-Ericksen equations when one allows for the Frank elasticity, (b) the molecular theory of the Leslie coefficients, (c) the interrelation between texture and dynamics, (d) the general form of the constitutive equations. We know virtually nothing about the last topic. Marrucci presents his view of the first three items, but clearly we are only just beginning to comprehend the complexities of nematorheology, and of the practical problems involved in achieving both molecular and macroscopic order in LPC's.

The reader might almost be tempted to conclude that at least in some ways our understanding of LPC's is rather impressive since it actually spans the quantumchemical to macroscopic levels. Still, one should be wary of the pitfalls inherent in a growing discipline. For one thing the number of really critical experiments attempting to disprove or verify current ideas is not high. In addition, few of the subdisciplines reviewed in this book have yet attained peaks in their development, so we should see quite a lot of progress in the coming decade. Personally, I am glad I did not heed the advice of several reputable scientists who, back in 1980, deemed polymer liquid crystals a field singularly devoid of merit. I hope the reader will agree with me that it is in fact an exciting field.

Delft, August 1990   T. Odijk

PART I

CONFORMATION

Conformational Analysis of Mesogenic Polymers

F. LAUPRÊTRE and C. NOËL

Laboratoire de Physicochimie Structurale et Macromoléculaire,
ESPCI, Paris, France

CONTENTS

1. INTRODUCTION

For low molecular weight liquid crystals (LMWLC's), the manner in which the molecules pack together in an ordered arrangement and the thermal stability range of the ordered arrangement depend on both the molecular structure and geometry (anisotropy, rigidity, linearity, planarity, etc.) of the mesogenic core and the equilibrium flexibility of the terminal substituents.[1-3] These two factors are of interest for both intra- and intermolecular interactions. The same factors must be taken into account when assessing the potential LC behavior of polymers.[4-9]

In spite of the rigidity of the molecules, liquid crystal polymers (LCP's) possess some conformational flexibility. The aromatic core often has internal degrees of freedom and flexible groups can give rise to different conformations. For example, internal rotations should be permitted (at least to some extent) about *p*-phenylene groups and C—C and C—O bonds in the following repeat unit:

The associated flexibility is of importance for the solubility characteristics of the polymer, chain-packing effects in the crystalline and liquid-crystalline states, and mechanical properties.

The first chapter of this book introduces the reader to the chemical structures of LCP's through their conformational characteristics. The primary aim is to present selected examples showing that calculations of intramolecular (conformational) energies can be carried out successfully to characterize and elucidate this type of conformational flexibility in LCP's. Of primary interest are the identification of the conformations of lowest energy and their comparison with known crystalline-state conformations of relevant model compounds. If a predicted conformation is nonplanar, the estimated energy difference between it and the planar geometry, and more generally the rotational energy profile, becomes of considerable importance. In a second step, calculations can be extended to take approximate account of intermolecular interactions, which one would expect to favor planar forms by their more efficient packing.

2. GENERAL CONCEPTS FOR CONFORMATIONAL ENERGY CALCULATIONS IN POLYMER CHAINS

2.1. Introduction

There are many advantages in a theoretical approach to conformational analysis, besides reproducing some data already attained by one of the many experimental methods. Most of the experimental techniques yield only a single piece of conformational information, usually relative to the minima of the potential or to their relative energy values, whereas conformational energy calculations can produce a complete potential function, then allowing the probability of molecular conformations between the actual conformers to be determined. Theory is of course appropriate for studying molecules that are not amenable to direct experimental investigation. Systematic computer calculations can be of great help in predicting the conformations of homologous compounds and in designing the chemical structures of candidates for specific applications.

Conformational energy determination of preferred conformations is based on the minimization of the total energy of the molecule, E_{tot}, with respect to all or some structural parameters. Even though different classifications of theoretical methods of conformational analysis exist, we have chosen as most illustrative the one proposed by Golebiewski and Parczewski[10] (Fig. 1). In this classification, in uniform methods E_{tot} is calculated by using only one theoretical scheme. Such an approach is provided by quantum mechanical methods in which all electrons, or at least all valence electrons, are considered.

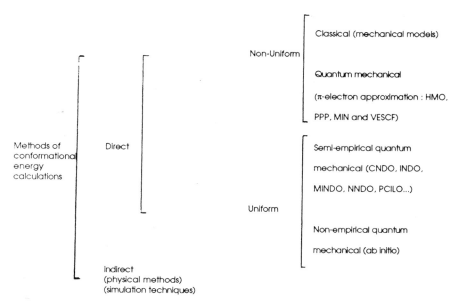

FIGURE 1
Classification of theoretical methods of conformational analysis.

In nonuniform methods, one assumes that:

$$E_{\text{tot}} = W + V \qquad (1)$$

where V and W correspond to the interactions of the nonbonded atoms and the remaining energy contributions, respectively.

2.2. Uniform Methods

Although the complete fundamental description of a molecule is provided by ab initio calculations, the uniform methods, which have proved useful in investigating the conformational properties of the constitutive units of LCP's, are mainly based on the concept of molecular orbitals (MO's) and the self-consistent field (SCF) theory.[11]

Each molecular orbital Ψ_s can be written as a linear combination of atomic orbitals (LCAO) φ_i:

$$\Psi_s = \sum_i C_{is} \varphi_i \qquad (2)$$

For closed-shell systems, the weighting coefficients, C_{is}, and the orbital energies of electrons, e_s, follow from the set of secular equations

$$\sum_j C_{js}(F_{ij} - e_s S_{ij}) = 0 \qquad j = 1, 2, \ldots, N \qquad (3)$$

where S_{ij} are the overlap integrals:

$$S_{ij} = \int \varphi_i \varphi_j \, dv \tag{4}$$

and the terms F_{ij} are the matrix elements of the Hartree-Fock operator:

$$F_{ij} = f_{ij} + \sum_{k,l} P_{kl} \left(G_{ij,kl} - \tfrac{1}{2} G_{il,kj} \right) \tag{5}$$

where

$$f_{ij} = \int \varphi_i^* (T + V_{\text{el-n}}) \varphi_j \, dv \tag{6}$$

$$P_{kl} = 2 \sum_{s=1}^{N/2} C_{ks}^* C_{ls} \tag{7}$$

$$G_{ij,kl} = \int \int \varphi_i^*(r_1) \varphi_j(r_1) \frac{e^2}{r_{12}} \varphi_k^*(r_2) \varphi_l(r_2) \, dv_1 \, dv_2 \tag{8}$$

T being the kinetic energy operator of an electron and $V_{\text{el-n}}$ the Coulomb interaction energy of the electron at a fixed position with the set of all nuclei. Since the Fock matrix, F, depends on the coefficients C, the equations are solved through an iterative procedure. An initial set of coefficients C is chosen and used to calculate the F matrix. Then, the equations are solved through a new set of coefficients. This process is repeated until consistency is attained; hence the name of self-consistent field method.

The total energy is given by the expression:

$$E = \sum_{s=1}^{N/2} e_s + \tfrac{1}{2} \sum_{k,l} P_{k_p} f_{k_p} + V_{\text{n-n}} \tag{9}$$

where the last term represents the Coulomb repulsion energy of nuclei or the atomic cores.

The Hartree-Fock method is based on an independent particle model, which allows for the possibility of finding two electrons in the same place at the same time; i.e., it neglects the correlation between the motions of electrons. A variety of methods has been used to obtain correlated wave functions. One popular approach is the configuration interaction (CI) method[12] in which it is assumed that the correct total wave function is a superposition of the wave functions of all possible configurations.

In the following, we confine our attention to semiempirical conformational analysis. Practical application of the SCF MO theory is limited by the capabilities of computers. To reduce the computer time and to extend the range of applications, numerical simplifications are introduced and integrals are often replaced by semiempirical parameters.

2.2.1. EXTENDED HÜCKEL THEORY (EHT)

The EHT, whose fundamentals we owe to Hoffmann,[13] assumes a generalization of the simple HMO approximation to include all valence electrons. In the EHT method pure AO's are used in place of hybrid orbitals. The overlap integrals S_{ij} are calculated theoretically for Slater-type atomic orbitals. The on-diagonal matrix elements, F_{ii}, are considered semiempirical parameters and taken as the negative values of the valence-state ionization potentials. Off-diagonal matrix elements are evaluated by means of the Wolfsberg-Helmholtz expression[14]:

$$F_{ij} = \tfrac{1}{2}\left(F_{ii} + F_{jj}\right)KS_{ij} \qquad (10)$$

where K is an empirical parameter. It is assumed that:

$$E \sim 2\sum_{s=1}^{N/2} e_s \qquad (11)$$

It must be noticed that the EHT method does not neglect the off-diagonal terms of the Fock matrix. Besides, it takes into consideration all valence electrons (σ and π electrons). In general, it can be said that the EHT method predicts results that, from a qualitative point of view are often acceptable. It should be pointed out, however, that the predictions are not always adequate, especially when atoms of different electronegativity exist in the molecule or when the size of the molecule becomes important.

2.2.2. COMPLETE NEGLECT OF DIFFERENTIAL OVERLAP (CNDO), INTERMEDIATE NEGLECT OF DIFFERENTIAL OVERLAP (INDO), MODIFIED INTERMEDIATE NEGLECT OF DIFFERENTIAL OVERLAP (MINDO), AND NEGLECT OF DIATOMIC DIFFERENTIAL OVERLAP (NDDO) METHODS

A detailed description of these methods is available.[11, 15] In these models, the Hartree-Fock elements, F_{ij}, are considered explicitly. However, numerical approximations are usually introduced. First of all, only valence electrons are considered explicitly. The nuclei and the inner shells are considered unpolarizable cores. The most important approximation, however, is introduced by the zero-differential overlap (ZDO) approximations: (1) in the CNDO method, $\varphi_i(r)\varphi_j(r) = 0$ except for $i = j$; (2) in the NDDO method, $\varphi_i(r)\varphi_j(r) = 0$ for $i \neq j$, provided φ_i and φ_j are the atomic orbitals of different atoms; (3) in the INDO method approximation, (1) is introduced in the case of integrals $G_{ij,kl}$, where orbitals φ_i, φ_j, φ_k, and φ_l refer to the same atom. For remaining integrals, approximation (2) is assumed.

Approximations (1)–(3) are not applied to the core integrals, f_{ij}. Many of the remaining integrals are estimated semiempirically, leading to various versions, such as CNDO/1, CNDO/2, MINDO (modified INDO), or MINDO/2.

To ensure the requirements of space invariance, additional approximations are required in the CNDO and INDO methods.

There are a considerable number of references concerned with the CNDO-type conformational analysis. Results obtained are in reasonable agreement with experiment. In this context the CNDO/BW method, developed by Boyd and Whitehead,[16] proves quite valuable for the study of conformational problems. It should be noted, however, that the CNDO/2 method fails to reproduce the experimental conformations of biphenyl[17] and stilbene,[18] which are acceptably reproduced by the EHT calculations.

In general, the INDO results do not afford an improvement over the CNDO/2 results in their agreement with experiment. It should be noted that the INDO method tends to overestimate the height of the barriers to rotation. In the MINDO method, the semiempirical parameters have been chosen in order to optimize the ground-state properties. Hence, the results obtained by the MINDO method are often more satisfying than those derived by using the CNDO and INDO methods.

Broadly speaking, better agreement with experiment is obtained when passing to more advanced treatments: EHT, CNDO, INDO, MINDO, NDDO.

2.2.3. PERTURBATIVE CONFIGURATION INTERACTION USING LOCALIZED ORBITALS (PCILO) METHOD

An intrinsic difficulty in the above described methods is the so-called electron correlation effect: the probability of finding an electron at a certain point in space is not affected by the fact that another electron can occupy the same space. The immediate consequence of the electron correlation effect is that the repulsion energies computed are overestimated. This shortcoming of the SCF–LCAO–MO scheme can be eliminated in different ways. As previously mentioned, the one most frequently used is the configuration interaction (CI) method.[12] An alternative treatment is provided by the PCILO scheme.[19-21] In this method one starts with a set of localized bonding and antibonding orbitals. In the next step one makes a perturbational treatment of configuration interaction to a rather high order.

This method has by far been that most abundantly used for computations in the field of biochemistry.

2.3. Nonuniform Methods

2.3.1. INTERACTION ENERGY BETWEEN NONBONDED ATOMS

The interaction energy, V, between the nonbonded atoms can be expressed as follows:

$$V = \sum V_j(r_j) \tag{12}$$

where the index j refers to the various pairs of nonbonded atoms and r_j is the distance between these atoms. The expression describes van der Waals interactions. The general features of interatomic potentials are very well known. At short distances, van der Waals interactions are repulsive. For a pair of atoms, the repulsion arises from the coulombic nuclear–nuclear interaction and the electron–electron coulombic overlap interactions summing up to a greater repulsive value than the attractive interaction between the nuclear core and electrons on the two different atoms. At large distances the curve is weakly attractive. The attraction is believed to be caused mostly by various types of dipole–dipole interactions. If neither of the interacting atoms has a permanent dipole moment, the primary forces are the so-called London dispersion interactions, which vary as r_j^{-6}. In practice, the pairwise potential interactions are normally represented by one of the two classical empirical functions: the Lennard-Jones potential function:

$$V_j(r_j) = -A_j/r_j^6 + B_j/r_j^{12} \tag{13}$$

or the Buckingham one:

$$V_j(r_j) = -A_j/r_j^6 + B_j \exp(-C_j r_j) \tag{14}$$

where A_j, B_j, and C_j are positive constants. In practice, these constants are treated as empirical parameters.[22]

The different nonuniform methods differ mainly in the way of estimating W and in the way they minimize the total energy. Among the contributions to the W term, depending on the molecule of interest, electrostatic and torsional interactions, deformation energies of the valence bonds and angles, hydrogen bonding, etc., have to be considered.

2.3.2. BOND ANGLE AND BOND LENGTH DISTORTION POTENTIAL FUNCTIONS

The energy needed to strain valence bond and bond angles can usually be accurately estimated from the knowledge of the "force constants" of the group of atoms within the molecule. The concept of a force constant implicitly assumes that valence bonds are essentially springs that obey Hooke's law. Under this assumption, the expression for the contribution to W can be written:

$$\Delta W = \sum_{i=1}^{N_a} \tfrac{1}{2}k_{a,i}\theta_i^2 + \sum_{i=1}^{N_b} \tfrac{1}{2}k_{b,i}S_i^2 \tag{15}$$

where N_a refers to all bond angles, N_b to all bond lengths, θ to the distortion angle, and S to the distortion length; $k_{a,i}$ and $k_{b,i}$ are the appropriate force constants.

2.3.3. TORSIONAL POTENTIAL FUNCTIONS

Changes in energy due to the torsion about a near single bond are given by:

$$\Delta W_{\text{rot}} = U_0(1 + \cos n\varphi) \tag{16}$$

where n is the periodicity of the function and φ is a generic angle of rotation. The barrier of rotation is $2U_0$.

Principles of the classical conformational analysis of organic molecules have been given by Hill[23] and Westheimer.[24] Other methods are basically extension of these. The reader who is interested in a detailed description of each of these mechanical models as well as their principal applications can consult the survey by Golebiewski and Parczewski.[10]

In principle, the above-described methods cannot be applied to systems that contain conjugated double bonds and exhibit large steric hindrance: a delocalization of local distortion has to be considered. However, Kitaygorodsky and Dashevsky[25] have shown that the above method remains valid in the case of small distortions. They have used this method successfully to predict the conformations of halogenobenzenes, halogenonaphthalenes, biphenyl, and binaphthyl.

An alternative approach was developed by Coulson and Senent.[26] According to these authors, the steric hindrance is counterbalanced by out-of-plane deformations of the conjugated π-electron system. Their method was improved by considering variation of bond lengths and valence angles and by replacing the hard-sphere model by a more sophisticated one.[27]

Classical conformational analysis is easily applicable even to relatively large molecules. There are, however, internal difficulties in the case of heteroatomic systems and π-electron systems with conjugated double bonds.

2.3.4. π-ELECTRON ENERGY IN CONJUGATED SYSTEMS

Hückel (HMO) Method.   The variation in total energy of a molecular conjugated system, due to a structural modification, is given by:

$$\Delta E = \Delta E_\pi + \Delta E_\sigma + \Delta V \tag{17}$$

The Hückel method is used to estimate ΔE_π. The total π-electron energy is here a sum of orbital energies of all π electrons. This is certainly a crude approximation, as the interelectronic repulsion energy is calculated twice. The structural variation can affect the molecular system in one of two ways, the first is to vary the bond lengths, preserving the molecular coplanarity; the second is where loss of the molecular coplanarity is involved.

In the first case, the modifications to be introduced in the calculations of ΔE_π are due to the variation of the resonance integrals, which are now expressed in the form of:

$$\beta_{ij} = K\beta_{ij}^0 \tag{18}$$

where β_{ij}^0 is the standard value of the resonance integral and K is a constant.

The loss of molecular coplanarity is due to the twisting of one or more bonds, the resonance integrals being expressed by:

$$\beta_{ij} = \beta_{ij}^0 \cos v_{ij} \tag{19a}$$

or

$$\beta_{ij} = \beta_{ij}^0 \cos^2 v_{ij} \tag{19b}$$

where v_{ij} is the twisting angle of the bond.[28–30]

Polansky[30] studied the molecules of biphenyl and of o- and p-terphenyl. He computed ΔE_π in the accustomed way and the remaining energy by the point charge model:

$$V = C \sum_{i<j} \sum 1/r_{ij} \tag{20}$$

where C is a constant. The barriers to rotation and the twist angles were satisfactory.

The Golebiewski group has made a great number of conformational studies using a self-consistent HMO method. In particular, they analyzed the conformations of neutral and ionic biphenyl[31] and cis- and trans-stilbene.[32] Their results were in good agreement with experiment.

Pariser-Parr-Pople (PPP) Method. This method was developed by Pariser and Parr[33] and was completed by Pople.[34] Again the validity of Eq. (17) is assumed. For closed-shell systems, the π-electron energy is given by:

$$E_\pi = 2 \sum_i^{\text{OCCMO}} e_i - \frac{1}{2} \sum_i^{\text{at.}} \sum_j^{\text{at.}} \left(q_i q_j - \frac{1}{2}P_{ij}^2\right)\gamma_{ij} \tag{21}$$

in which the first summation represents the sum of the orbital energies of all the π electrons and the second term is a correction due to the energy of π-electron interaction; q_i is the mean number of π electrons at the atom i; P_{ij} is the mobile bond order between atoms i and j; and γ_{ij} is the Coulomb integral between two $2p_\pi$ electrons, one being at atom i and the other at atom j.

In many cases PPP-based conformational analysis yields rather satisfactory results. Fisher-Hjalmars[35] obtained good agreement with experiment for the barrier to internal rotation of biphenyl. Dewar and Harget,[36] using

simultaneously the Hückel σ-, Hückel π-, and Pople π-electron approximations, obtained $\theta = 40°$ for the dihedral angle of biphenyl, in excellent
agreement with the experimental $\theta = 42°$.

2.3.5. OTHER CONTRIBUTIONS TO W

Among the other contributions that must be introduced to represent the
energy function of a given molecule in a given state, one may also have to
take into account interactions between electric dipoles or charges, hydrogen
bonding, interactions with the solvent or with other molecules. Description of
a number of appropriate functions are available.[37]

2.4. Application to Polymer Structure

It is difficult to describe precisely macromolecular geometries because of the
immense size of these molecules and the large number of possible degrees of
structural freedom. One popular approach is to use several of the earlier
described methods simultaneously on judiciously selected low molar mass
compounds that model the fragment under study. The method used must be
chosen with full awareness of its reliability in treating the property or
combination of properties being explored in a given calculation. Of fundamental importance in being able to predict molecular structure is the avail-
ability of a reliable set of values for bond lengths and bond angles. These
geometrical parameters are usually estimated from x-ray crystallographic
data of related compounds.

Almost all conformational energy calculations deal with the conformation of isolated molecules, thus ignoring the possible effect of the environ-
ment of the system. Calculations yield conformational energy maps that give
the positions of the energy minima, i.e., the nature of the preferred conformations, and the heights of the energy barriers. From these conformational
energy maps, static properties of the isolated macromolecule depending on
the conformation can be derived such as ^3J coupling constants and Nuclear
Magnetic Resonance (NMR) chemical shifts. Of particular interest is the
chain statistics approach described by Volkenstein[38] and Flory,[39] which
allows the calculation of end-to-end distances and persistence vectors. The
persistence vector $\langle \mathbf{R} \rangle$ is defined as the average of the end-to-end vector over
all the conformations of the chain consisting of n units. It is obtained by
taking the conformational averages of the serial products of transformations:

$$\mathbf{R} = \mathbf{I}_1 + \sum_{k=1}^{n-1} \prod_{i=1}^{k} \mathbf{T}_i \mathbf{I}_{k+1} \qquad (22)$$

where the $\mathbf{T}_i$ matrix is the matrix of transformation of a vector in reference
frame $x_{i+1} y_{i+1} z_{i+1}$, associated with the $i + 1$ unit, $\mathbf{I}_{i+1}$, to its representation
in reference frame $x_i y_i z_i$. The mean-square end-to-end distance $\langle R^2 \rangle$ can

be obtained by a similar matricial calculation. Such chain statistics approaches have been successfully applied to a number of polymers by Tadokoro and co-workers[40, 41] and by Flory and his co-workers.[42, 43]

Although the different methods used to calculate conformational energies do not directly simulate intramolecular motions, they allow, by comparing the heights of the energy barriers of two or more conformations, derivation of some conclusions concerning local motions observed, for example, by spectroscopic investigations of local dynamics. Such an approach may also aid in identifying the motional processes that are responsible for the observed secondary relaxation processes. Very recent techniques now include molecular dynamics simulations.[44, 45]

While such calculations on isolated molecules are necessary and convenient as points of departure, further studies may be needed to describe the solution, melt, or crystalline state, in which intermolecular forces compete with the intramolecular forces.

In solution, interactions of the solute molecule with the solvent may be either of steric and coulombic type only, or they may also involve hydrogen bonding. A very simple manner of accounting for steric interactions has been proposed by Sundararajan et al.[46] for polymer chains with large and planar R side groups. When the distance between the side groups is such that it allows the penetration of a solvent molecule, the R $\cdots$ R van der Waals interaction is replaced by a R $\cdots$ solvent interaction. In the case of hydrogen bonding, one way to attack the problem is to calculate the conformational energy map of the hydrogen-bonded complex. Such an approach may lead to a deeper understanding of the extent to which the intermolecular interactions are modified by protonation of the chains, which occurs, for example, in the strongly acidic media used as solvents for some lyotropic polymers.

The traditional approach to the evaluation of packing energy in crystals[47, 48] rests on the separation of intramolecular interactions from intermolecular ones. The conformation of the molecule is first refined to minimize all intramolecular energy contributions and subsequently a rigid-body adjustment in the crystal lattice is carried out to minimize intermolecular interactions. This two-step procedure may, however, bias the final results in a significant way and further refinements of the intramolecular structure in the presence of intermolecular interactions may be needed. For example, Kusanagi et al.[49] applied this two-step procedure to poly(ethylene oxybenzoate). They found that a calculated isolated chain structure was far from one allowing solution of the x-ray structure. By minimizing the packing energy with chain torsional angles included, they were able to solve the x-ray structure.

A more realistic single-step approach, where both intra- and intermolecular interactions are simultaneously taken into account, is possible provided that a small portion of the crystal is taken into account. Such a computer program has been successfully employed in the case of isotactic *trans*-1,4-poly(1,3-pentadiene)[50] and α- and γ-polypivalolactone.[51] It yielded results in good agreement with the models directly refined from the powder x-ray diffraction profile and from electron diffraction data. The conformational differences displayed by the polymer chains in the crystalline forms are reproduced and can be explained in terms of different packing interactions.

A similar methodology has also been proposed by Sorensen et al.[52] Their procedure was based on the desire to be able to calculate a variety of useful crystal properties in accompaniment with the packing prediction. These properties include vibrational analysis to determine dispersion curves, infrared and Raman spectra, thermodynamic functions, and calculation of elastic constants. Sorensen et al.[53] applied this method to two examples. The first was polyethylene, chosen as a standard because its structure and properties are well known. The latter example was polyoxymethylene, which crystallizes in two forms. Good agreement between calculated and experimental structures was found.

3. CONFORMATIONAL ANALYSIS OF SOME THERMOTROPIC SYSTEMS

3.1. Introduction

Thermotropic mesogens are those compounds which exhibit a mesophase in the molten state. Traditionally two major classes of thermotropic liquid-crystalline polymers have been identified: the so-called mainchain and sidechain types (LCMP's and LCSP's, respectively) (Fig. 2). More recently other variants have appeared: these are combined LCP's,[54,55] which are hybrid between LCMP's and LCSP's, and the rigid-rod types described by Watanabe et al.[56] A great wealth of literature already exists in the form of unified texts and reviews that detail both the major classes of LCP's.[57-66] Bibliographic data have been compiled[67] and reviews more or less specific to mainchain[68-71] or comb[72-75] polymer systems have appeared. We shall be concerned here only with LCMP's, for simplicity indicated as LCP's.

Thermotropic LCP's include polymers such as cellulose derivatives and poly(n-hexyl isocyanate), which also exhibit lyotropic behavior. However, the most common units in thermotropic LCP's are benzene rings interlinked at para positions:

However, such rodlike molecules tend to be infusible, largely intractable crystalline solids. Liquid-crystalline properties are observed for oligomers of poly(p-phenylene) with $n = 5$ to 7 but when $n > 7$ decomposition occurs

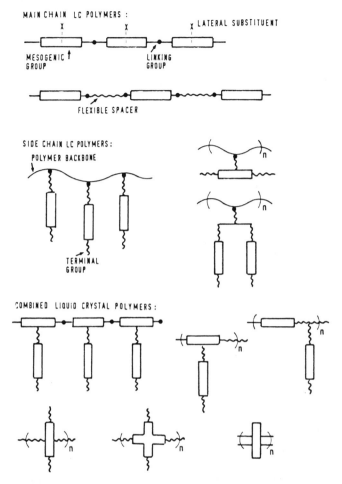

FIGURE 2
Typical structural modifications for LCP's.

below the melting points.[76] Melting points of 610 and 600°C were measured by Differential Scanning Calorimetry (DSC) (scan rate of 80° min^{-1} to minimize degradation) for poly(p-hydroxybenzoïc acid) and poly(p-phenylene terephthalate), respectively.[77] Hence, the problem of thermotropic LCP design is to lower the melting point to a melt processable range without destroying liquid crystal formation.

There are four basic methods that can be used to reduce the transition temperatures of LCP's by disrupting the perfect regularity of simple but intractable para-linked aromatic polymers[60–63, 68, 71, 77]: (1) Introduction of flexible segments to separate mesogenic groups placed in the mainchain from each other, (2) introduction of rigid kinks into the straight polymer chains,

(3) substitution of the aromatic rings, and (4) addition of an element of asymmetry to the mainchain by copolymerizing mesogenic units of different shapes.

The following scheme illustrates the two cases of LCP's incorporating a rigid kink and of segmented LCP's:

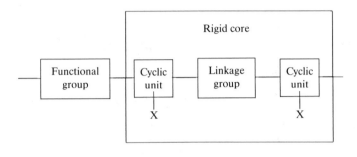

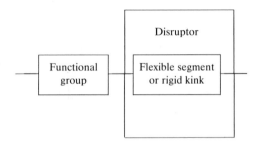

The rigid core consists of at least two aromatic (or more rarely cycloaliphatic and/or heteroaromatic) rings connected in the para positions by short rigid links. The structures of typical LCP's can include as linkages between aromatic rings, aside from direct attachment (such as in biphenyl and terphenyl units) the imino, azo, azoxy, *trans*-vinylene (or *trans*-stilbene), ester and methylene groups. As discussed above substituents such as F, Cl, Br, CH_3, OCH_3, phenyl, or n-alkyl group may be introduced into the mesogenic units. Other substituents may of course be employed. Examples are highly polar groups (e.g., $-CN$, $-NO_2$). Meta- or ortho-aromatic isomers and 2,6-disubstituted naphthalene derivatives may be incorporated as rigid kinks to offset the linearity of the rigid main chain. The most commonly used connecting units joining rigid core to disruptor are ester and ether groups. More recent studies now include the use of amide, imine, methane, and carbonate groups. The flexible segments are usually sequences of the polymethylene type. Also commonly investigated as spacers are poly(ethylene oxide) and polysiloxane chains.

Typical LCP's are neither rigid nor highly symmetrical; the aromatic core often has low symmetry and internal degrees of freedom, and the flexible segments can give rise to different conformations. Since from Fourier transform infra-red (FTIR) spectroscopy, electron spin resonance (ESR), and NMR investigations the thermal behavior of these polymers appears to be a sequence of changes from the solid state to the isotropic liquid phase, gradually liberating one degree of freedom after another, this flexibility may be important in this regard. The present section is not intended to be an updated comprehensive collection of conformational characteristics of the mesogenic moieties and disruptors currently used to synthesize LCP's. Some examples have been chosen to show how conformational analysis can be used to predict the way in which changes in molecular structure affect the properties of LC's induced by heating LCP's.

3.2. Mesogenic Core

A number of theoretical calculations have been devoted to LCP's, but most of them have been applied to the chemical units listed in Table 1. Their main first objective was to evaluate conformational energy maps, indicating the molecular energy as a function of the internal rotation angles.

PCILO calculations were carried out to obtain conformational energy profiles of molecules **1–5**[78]:

Table 1
Common Thermotropic Systems

Rigid core			Disruptor	
Cyclic unit	Linkage group	Functional group	Flexible segment	Rigid kink
(1,4-phenylene ring)	$-O-$	$-O-$	$+CH_2+_n$	(naphthalene)
(p-phenylene, $n = 2, 3$)	$-\overset{O}{\overset{\|}{C}}-O-$	$-\overset{O}{\overset{\|}{C}}-O-$	$+CH_2-CH_2-O+_n$	(1,2-disubstituted benzene)
(biphenyl, X ... X substituents)	$-CH=CH-$	$\overset{O}{\overset{\|}{-O-C-}}$		(1,3-disubstituted benzene)
	$-N=N-$	$-O-\overset{O}{\overset{\|}{C}}-O-$		(diphenyl ether, $-O-$)
	$-N=C-$			
	$-N=N-\!\!\to\!\!O$			

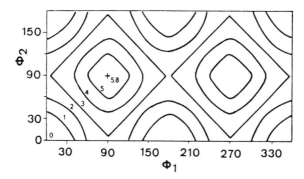

FIGURE 3
Conformational energy map (ϕ_1, ϕ_2) of methylterephthalate.
(Reproduced with permission from Ref. 78. Copyright Gordon &
Breach Science Publ.)

The internal rotation angles ϕ_i define the orientation of an aromatic ring
with respect to the adjacent carboxyl group. The angles Ψ_i characterize the
relative orientation of two adjacent rings.

The conformational energy map of methyl terephthalate (**1**) is shown in
Fig. 3. Two energy minima are found at $\phi_1 = \phi_2 = 0°$ and $\phi_1 = 0°$; $\phi_2 = 180°$.
The former corresponds to the cis configuration of the carboxyl groups
relative to the central ring and the latter coincides with the trans configuration. Coplanarity of the carboxyl group with the benzene ring guarantees
maximum overlapping of electrons of the participating atoms. The height of
the energy barrier is 2.9 kcal mol^{-1}. The energy maximum is obtained when
$\phi_1 = \phi_2 = 90°$. Besides, ϕ_1 and ϕ_2 rotations are independent.

For comparison it is useful to cite earlier empirical calculations carried
out by Hummel and Flory[42] and by Tonelli[79] in an attempt to determine the
preferred angles and evaluate the energy barriers for methylbenzoate and
terephthalate moiety, respectively. Very similar potential energy maps were
obtained. The equilibrium conformations also correspond to the planar cis
and trans configurations. Only the height of the energy barrier differs:
5 kcal mol^{-1} and 3 kcal mol^{-1} according to Hummel and Flory[42] and
Tonelli,[79] respectively. Such discrepancies may be due to the difficulty in
evaluating conjugation energy in empirical calculations.

These results are in reasonable agreement with experimental observa-
tions on dimethyl terephthalate. Indeed, the energy difference between the
cis and trans isomers was found to be 0.05 kcal mol^{-1} from dipole moment
measurements in solution.[80] Besides, infrared and Raman studies[81] indicated
that the populations of the cis and trans conformers are nearly the same in
the melt. Finally, the molecular conformation in the crystalline state was
found to be the trans one, the phenyl ring standing at an angle of 4.7° to the
plane of the carboxyl group.[82] The nearly planar conformation of the terephthaloyl residue favors the molecular packing in the crystal and enhances

the attractive intermolecular interactions between the ester groups of neighboring molecules.

A prominent feature of the preceding results is that the virtual bond spanning the benzene ring and connecting the carbonyl carbons should behave as a statistical freely rotating link[39] in polyesters based on terephthalic acid. This model of the terephthaloyl residue yields chain dimensions in substantial accord with the values deduced from experiments.[83]

The effect of reversing ester groups should be mentioned here (compare **1** and **6**, below). Conformational energies have been calculated for phenylacetate. An empirical force field (6-exp type) supplemented by terms for frame distortion and electron delocalization was used for this purpose.[42] Bond angles and bond lengths were adjusted to values that minimize the total energy at each value of the torsion angle. Conformational energy calculated for this molecule exhibits maximum at coplanarity of the ester group with phenyl, owing to steric repulsions involving orthohydrogens. The stable conformations of phenyl acetate are those in which the plane of the ester group is rotated $58 \pm 10°$ from the phenyl plane. Accordingly, four conformations representing combination of rotations ϕ_1 and $\phi_2 = 58 \pm 10°$ about the phenylene axis are accessible to the p-diacetoxy benzene[80]:

$$CH_3-\overset{\overset{\textstyle O}{\|}}{C}-O \overset{}{\underset{\phi_1}{\longrightarrow}} \langle\bigcirc\rangle \overset{}{\underset{\phi_2}{\longrightarrow}} O-\overset{\overset{\textstyle O}{\|}}{C}-CH_3$$

6

For biphenyl (**2**), values of the conformational energy, taken relative to that for the coplanar form, are plotted versus Ψ_1 in Fig. 4. The PCILO calculations give an absolute conformational energy minimum at $\Psi_1 = 40°$, which is in good agreement with the torsional angle of approximately 42–45° between the planes of the rings in the vapor state.[84] The height of the energy barrier is 1.9 kcal mol^{-1}. It is worth noting that the dihedral angle Ψ_1 lies in the range 20–25° in solution[85, 86] and in the melt,[87] while biphenyl is planar or nearly so in the crystalline state[88–92] at room temperature. This planar conformation, however, has been claimed[91, 93] to be the result of a statistically centered arrangement with large oscillations around the inter-ring bond. The coplanar conformation is hindered by steric interferences involving the ortho hydrogen. However, in the crystalline state intermolecular interactions favor the coplanar form by its more efficient packing and, in so doing, counteract the intramolecular steric hindrance, especially since the barriers to planarity are relatively low.[94]

Bastiansen and Samdal[95] and Häfelinger and Regelmann[96] summarized the theoretical calculations of different degrees of sophistication carried out on biphenyl. Calculations by molecular mechanics[97–100] with separate consid-